MW00462683

Congratulations!
You are in the process of getting SUPER SMART in the topic of Neurology!

Notice the **color coded** words & <u>**sound it out sections**</u> throughout...
You will learn complex, advanced vocabulary!
Review the words as you read to help the material stay fresh.

Neurology: The Amazing Central Nervous System is PACKED with info!
If necessary, read this AWESOME book section by section, or if you are
ready to learn a TON of cool new info, READ THE WHOLE THING AT ONCE!

I personally love to draw and doodle so I included extra pages in the
back for your mind to create! Enjoy

Share this book with your friends. Help them become SUPER SMART too!

Featured on page two is the artwork of 9 year old John Clint Lawhon who
won the Art Challenge Competition. Thank you for sharing with us your
concept of the Central Nervous System!

-April Chloe Terrazas

this SUPER awesome Book Belongs to:

_____

Winner of the Neurology Art Competition: John Clint Lawhon!

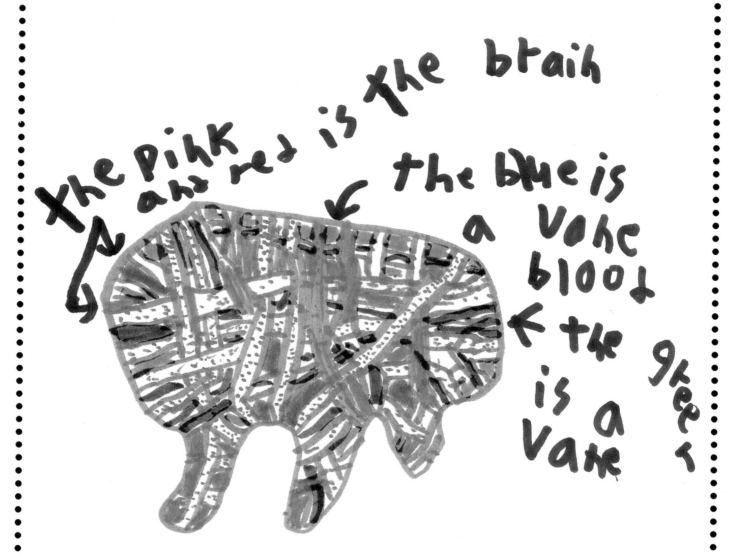

the pink and red is the brain

the blue is a vane blood

the green is a vane

John Clint Lawhon

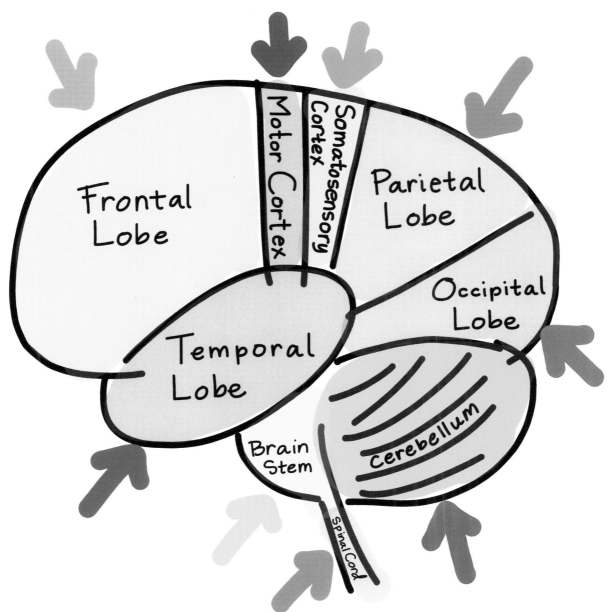

Neurology: the Amazing Central Nervous System

Written and Illustrated by: APRIL CHLOE TERRAZAS

Dedicated to:

Granny Emma Marie

Thank You for your support!!! ☺

Neurology: The Amazing Central Nervous System. April Chloe Terrazas, BS University of Texas at Austin.
Copyright © 2013 Crazy Brainz, LLC

Visit us on the web! www.Crazy-Brainz.com

Cover design, illustrations and text by: April Chloe Terrazas

The Central Nervous System is...

Sen-trul
Ner-vus
Sis-tem

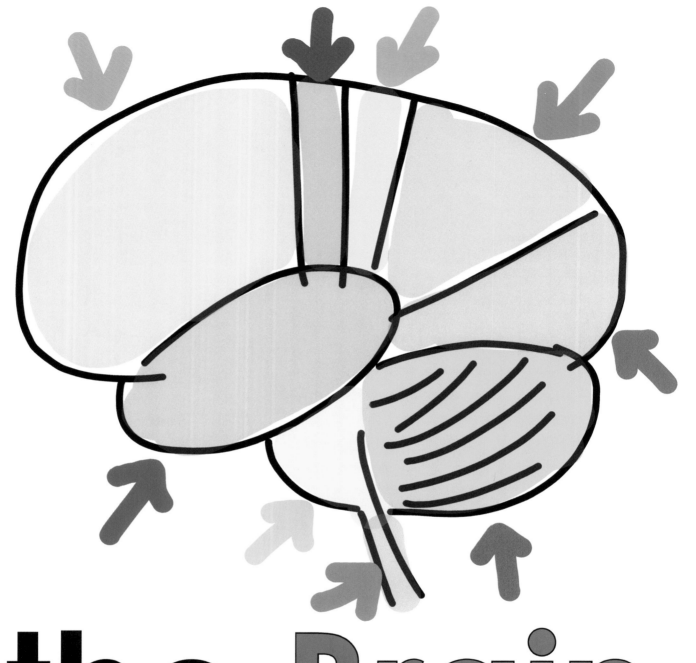

the Brain,

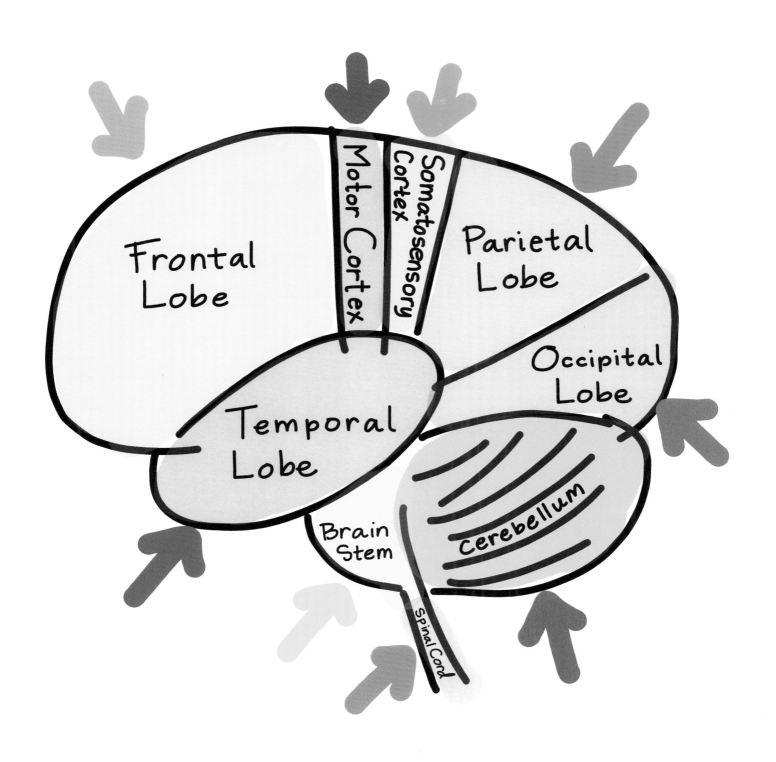

What do you think each part of the brain does?

Are YOU ready to

learn about the the BRAIN ???

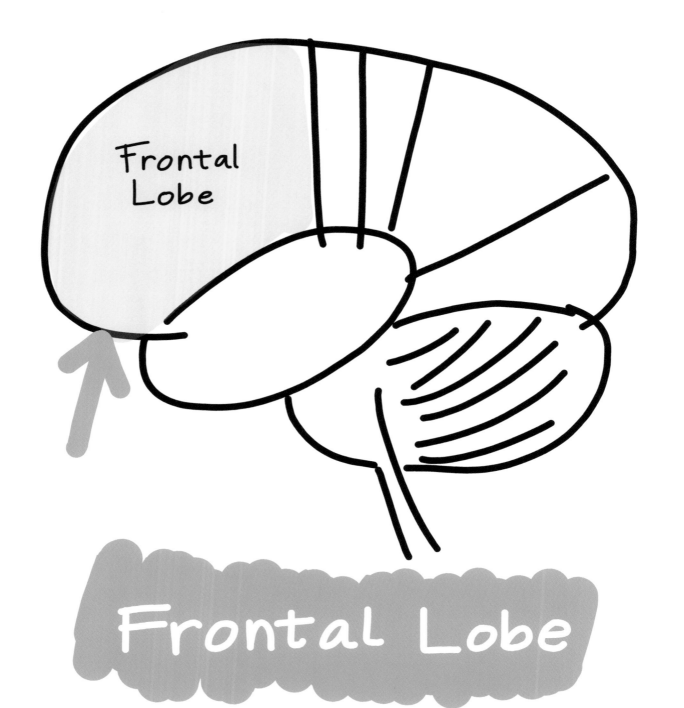

Frontal Lobe

Frontal Lobe

Touch your forehead.

The Frontal Lobe is the part of your brain at your forehead.

The **Frontal Lobe**
<u>**makes you speak and move**</u>.

Say "Hello. How are you today?"
Move your arms like a bird.

That is your Frontal Lobe working!

**The Frontal Lobe also helps you
solve problems like this one:
100 + 100 = ?**

**Can you think of another
problem that your
Frontal Lobe helps you solve?**

EXCELLENT WORK!

Remember this:
Frontal = Forehead

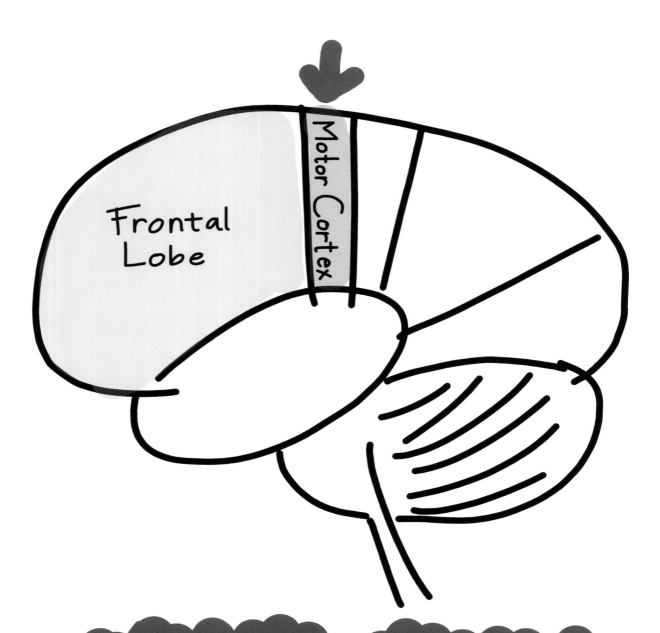

Motor Cortex

Sound it Out	Sound it Out
1. MO	1. KOR
2. TR	2. TEX

The Motor Cortex
controls voluntary movement.

Voluntary movements are any movements that you <u>think about</u> doing, *for example*:

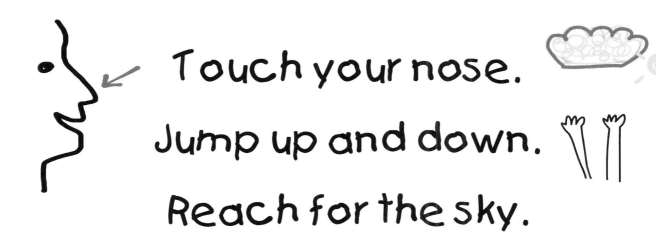

Touch your nose.

Jump up and down.

Reach for the sky.

That is your **Motor Cortex** working!

What other movements do you think your Motor Cortex controls?

Remember this:
Motor = Move

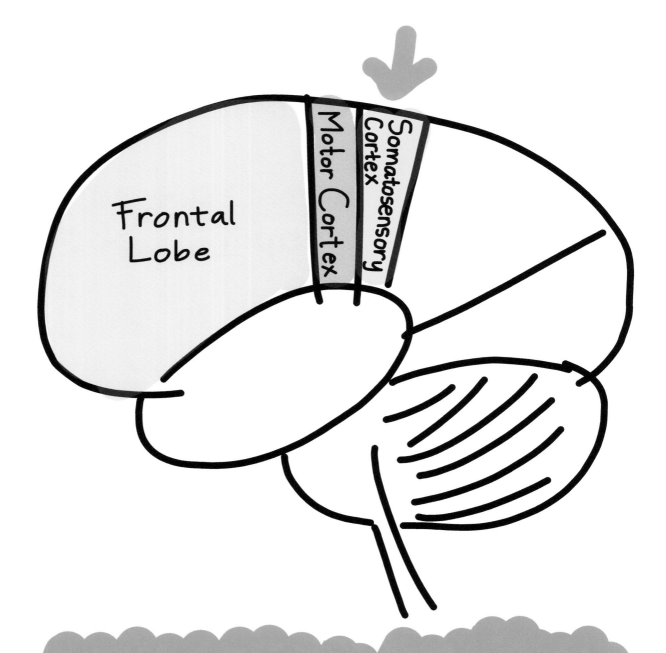

Frontal Lobe

Motor Cortex

Somatosensory Cortex

Somatosensory Cortex

Sound it Out	Sound it Out	Sound it Out
1. SO	4. SEN	1. KOR
2. MA	5. SOR	2. TEX
3. TO	6. EE	

The Somatosensory Cortex <u>receives</u> <u>signals</u> sent from all over your body.

Signals can be sent from your eyes, ears, mouth and nose!

Do you smell flowers?
Do you see the bright blue sky?
Do you taste yummy food?

The signals from all of these are sent to the Somatosensory Cortex of your brain!

What other signals do you think your Somatosensory Cortex receives?

Remember this:
Somatosensory = Signal

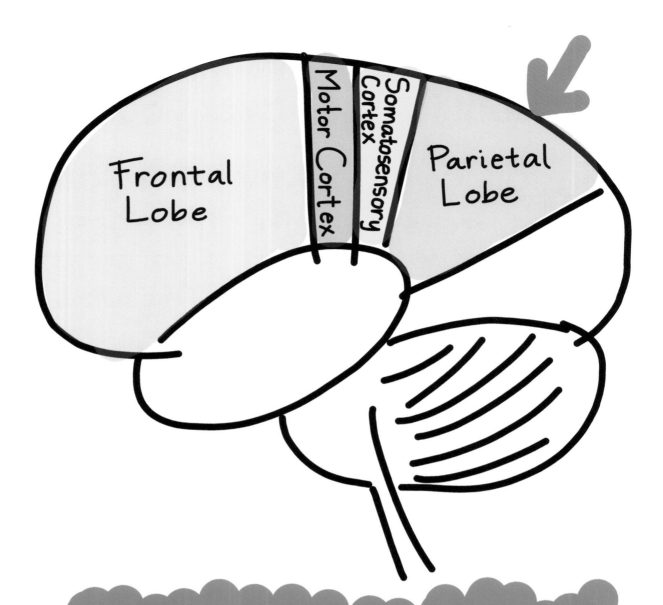

Parietal Lobe

Sound it Out

1. PA
2. RI
3. EH
4. TUL

Sound it Out

1. LOB

The **Parietal Lobe** <u>senses</u> things like touch, temperature and pain.

Do you know what paper feels like? Can you feel the heat from the sun?

You are able to <u>sense</u> these things because the **Parietal Lobe** of your **brain** is hard at work.

What other sensations do you think your **Parietal Lobe** processes?

Remember this:
Parietal = Sensation

Review the sections of the brain you have learned so far.

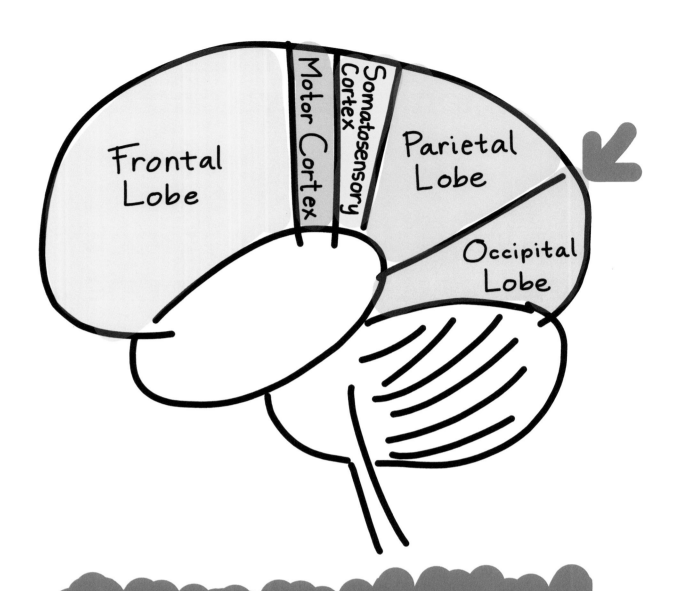

Occipital Lobe

The Occipital Lobe <u>processes</u> the signals from your eyes so you <u>understand</u> what you are seeing.

Do you see the difference between Pink and Blue?

Do you know what a dog looks like?

You can understand what you are seeing because of the Occipital Lobe of your brain!

What other signals do you think your Occipital Lobe helps you understand?

Remember this:
Occipital = Understand

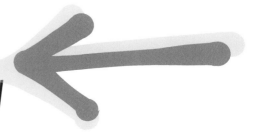

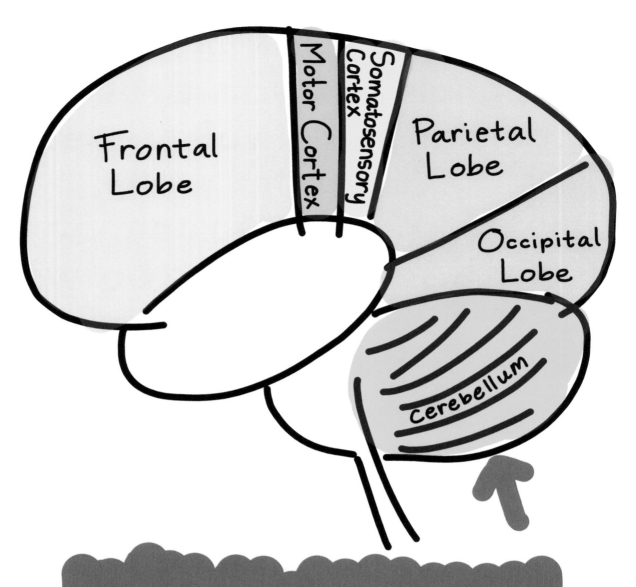

Frontal Lobe

Motor Cortex

Somatosensory Cortex

Parietal Lobe

Occipital Lobe

Cerebellum

Cerebellum

Touch the back of your head by your neck.

This is where your Cerebellum is located.

The Cerebellum controls balance, movement and coordination (how your muscles all work together).

Do you know how to ride a bike?

Have you ever seen a surfer on a big wave?

The Cerebellum controls balance so you can ride a bike and surf!

What other activities do you think your Cerebellum controls?

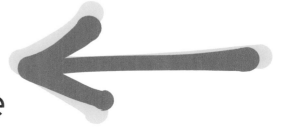

Remember this:
Cerebellum = Balance

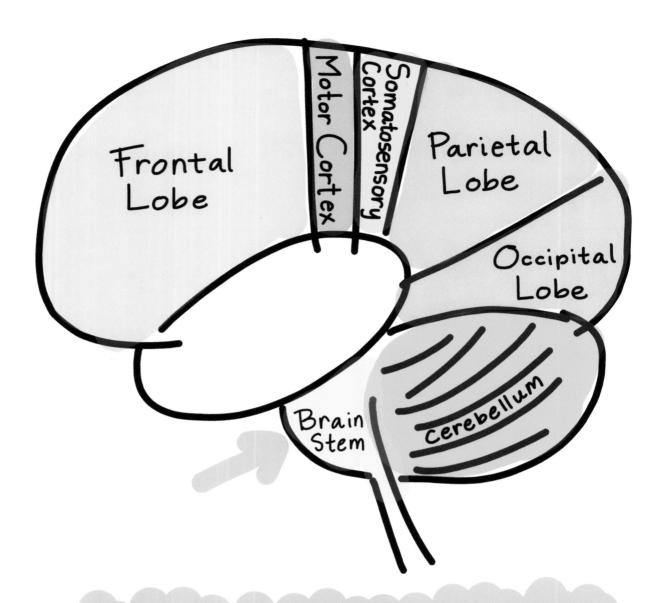

Frontal Lobe

Motor Cortex

Somatosensory Cortex

Parietal Lobe

Occipital Lobe

Brain Stem

Cerebellum

Brain Stem

Sound it Out

1. BRANE

Sound it Out

1. STEM

The Brain Stem connects the brain to the spinal cord.

The Brain Stem <u>controls heartbeat</u>, breathing, digestion and waking you up in the morning!

Do you have to think about breathing or making your heart beat?

No!

Your Brain Stem controls that action for you.

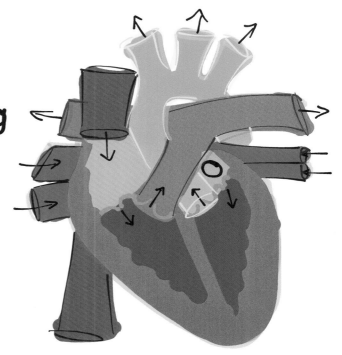

THE HEART
The arrows show the direction of blood flow in and out of the heart. Learn more about the heart in *Cardiology, book 7 of the Super Smart Science Series.*™

Remember this:
Brain Stem = Heartbeat

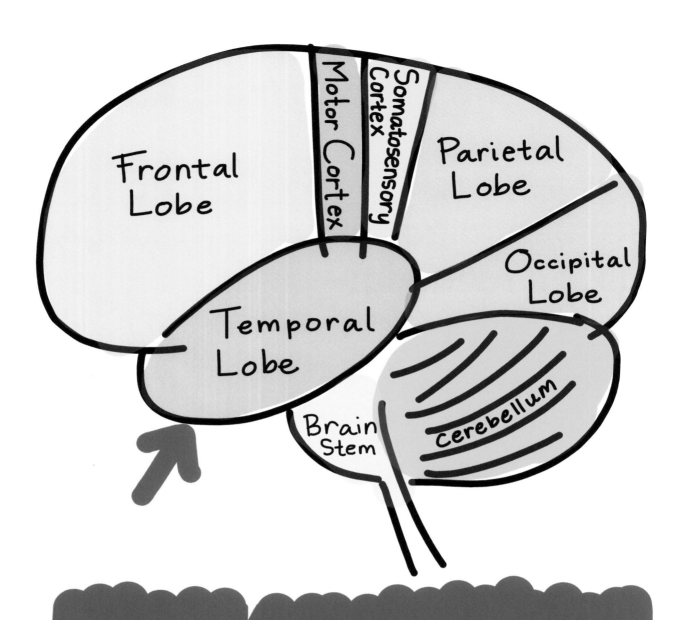

Temporal Lobe

Sound it Out

1. TEM
2. POR
3. UL

Sound it Out

1. LOB

The Temporal Lobe <u>controls hearing</u>. It receives sounds and speech and comprehends what you are hearing.

Can you hear a bird chirp?

Do you recognize that sound as being a bird chirp?

tweet! tweet!

You can hear and understand what you are hearing because of the Temporal Lobe of your brain.

What other things does your Temporal Lobe allow you to hear?

Remember this:
Temporal = Ears

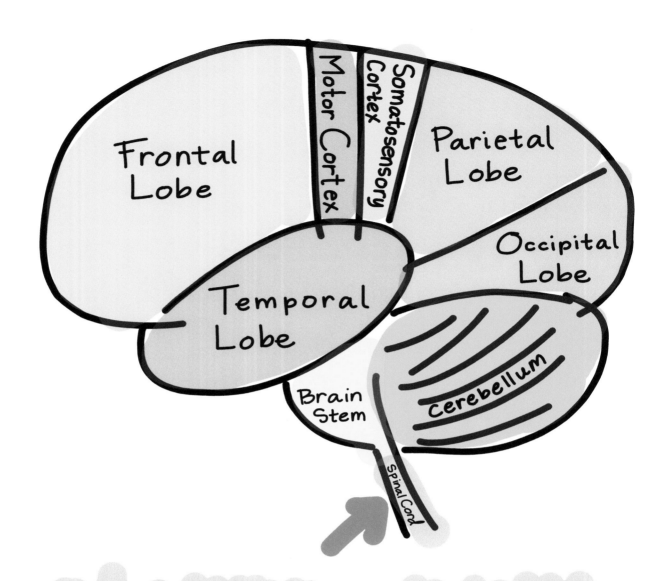

Spinal Cord

The Central Nervous System is the
BRAIN + SPINAL CORD

The spinal cord starts at the base of
the brain next to the brain stem
and continues all the way
down your back.

The spinal cord is the pathway for
messages between the
brain and body.

The spinal cord is protected by
bumpy bones called vertebrae.

Vertebrae

Sound it Out
1. VER
2. TE
3. BRAY

You have:

8 cervical vertebrae,

Sound it out: SER-VI-KUL

12 thoracic vertebrae,

Sound it out: THOR-AS-IK

5 lumbar vertebrae,

Sound it out: LUM-BAR

5 sacral vertebrae

Sound it out: SA-KRUL

and

1 coccygeal vertebra!

Sound it out: KOX-E-JE-UL

CERVICAL 8

THORACIC 12

LUMBAR 5

SACRAL 5

COCCYGEAL 1

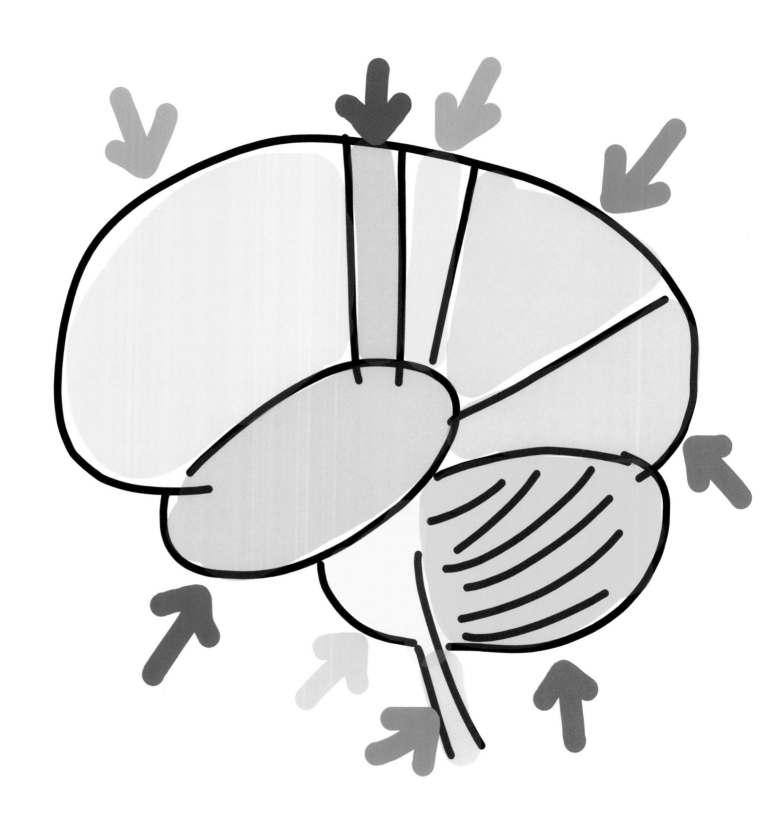

Match each colored section in the picture to the correct name.

Frontal Lobe
Motor Cortex
Somatosensory Cortex
Parietal Lobe
Occipital Lobe
Cerebellum
Brain Stem
Temporal Lobe
Spinal Cord

Do you remember the function of each part of the brain?

GOOD JOB!

Who wants to learn more?!

**Remember, the
Central Nervous System is the
BRAIN + *SPINAL CORD*.**

What is the
spinal cord
made of?

Bundles of Neurons!

Neuron

Sound it Out

1. NUR
2. ON

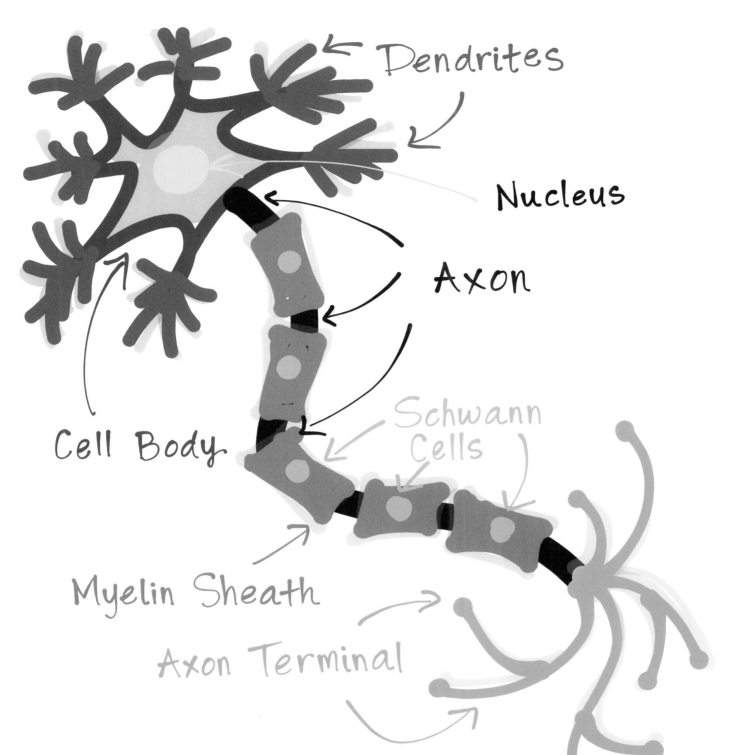

Dendrites

Nucleus

Axon

Cell Body

Schwann Cells

Myelin Sheath

Axon Terminal

This is a neuron. Neurons send messages across the body. What is the function of each part of the neuron? Let's see!

Nucleus

Nucleus

The nucleus contains DNA.

The nucleus is located inside something... inside what?

Turn the page and find out!

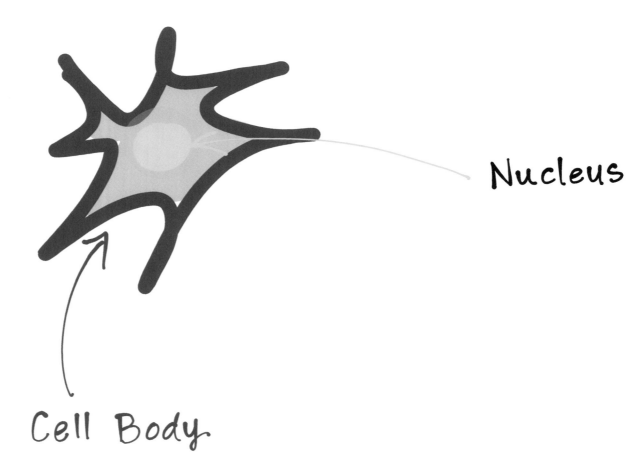

Nucleus

Cell Body

Cell Body

The nucleus
is located inside
the cell body!

The cell body receives
a message from a part
of the neuron, then
sends it to another part.

Where is this message
coming from?
And where does it go?

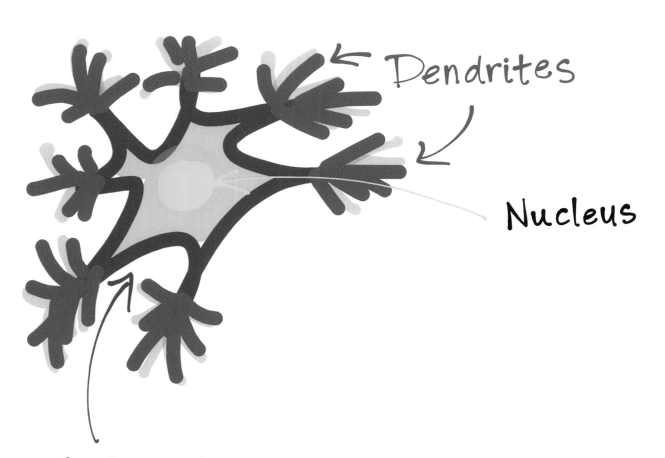

Dendrites

Nucleus

Cell Body

Dendrites

The message comes from the **dendrites**.

The **dendrites** send the message to the **cell body**.

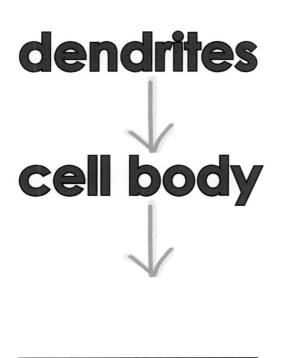

dendrites

cell body

_____

Where does the message go from there?

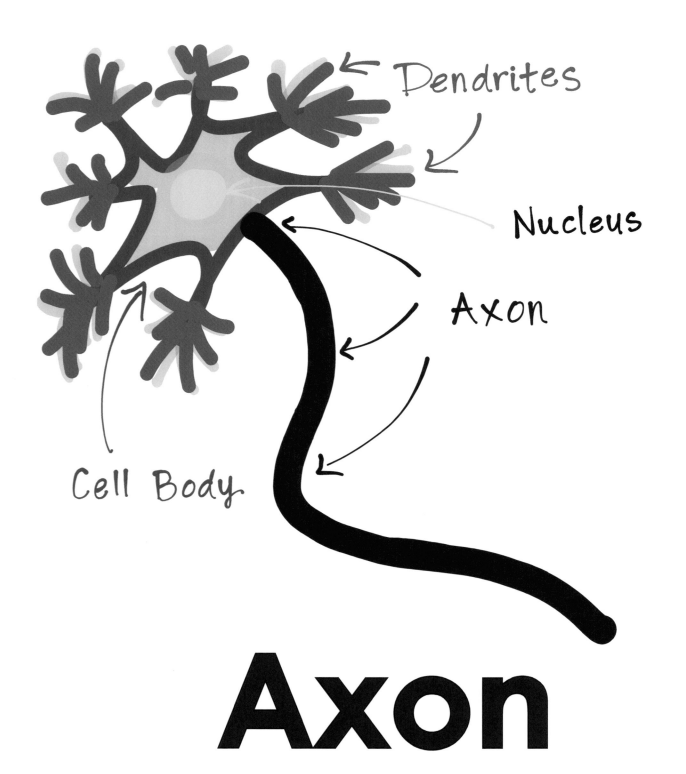

Dendrites

Nucleus

Axon

Cell Body

Axon

Sound it Out
1. AX
2. ON

The message leaves the **cell body** and goes down the **axon**.

Review the path of the message so far:

dendrites

↓

cell body

↓

axon

↓

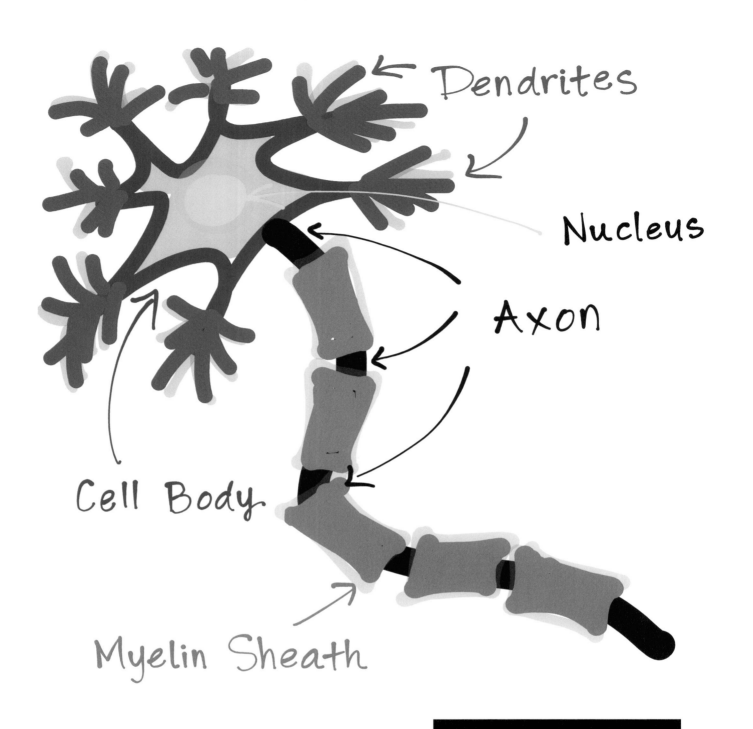

Dendrites

Nucleus

Axon

Cell Body

Myelin Sheath

Myelin Sheath

Sound it Out
1. MY
2. LIN
3. SHEETH

The myelin sheath wraps around the axon to help the message send quicker.

The myelin sheath does not cover the entire axon.

Do you see the axon between the myelin sheaths?

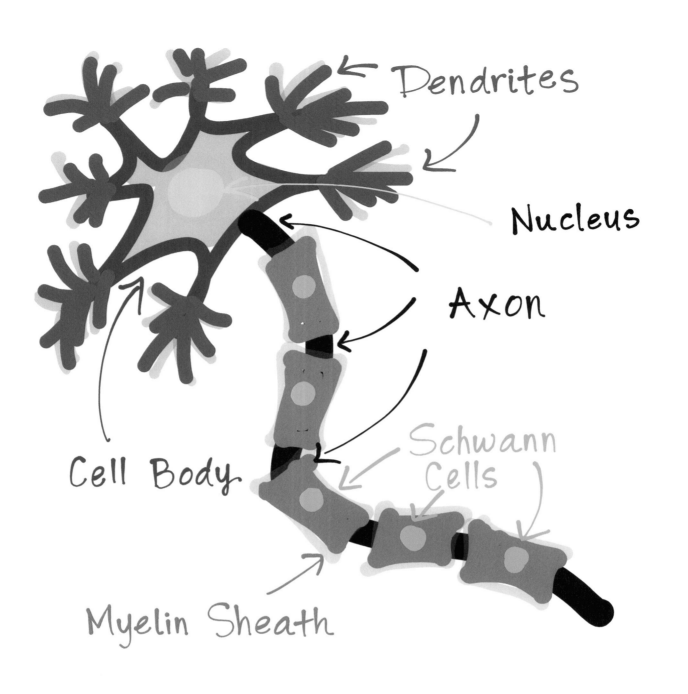

Dendrites

Nucleus

Axon

Cell Body

Schwann Cells

Myelin Sheath

Schwann Cells

Schwann cells **make**
the myelin sheaths that
wrap around the **axon**.

Do you remember what
the myelin sheath does?

What is the path of the
message from the
beginning?

_____ → _____ → _____

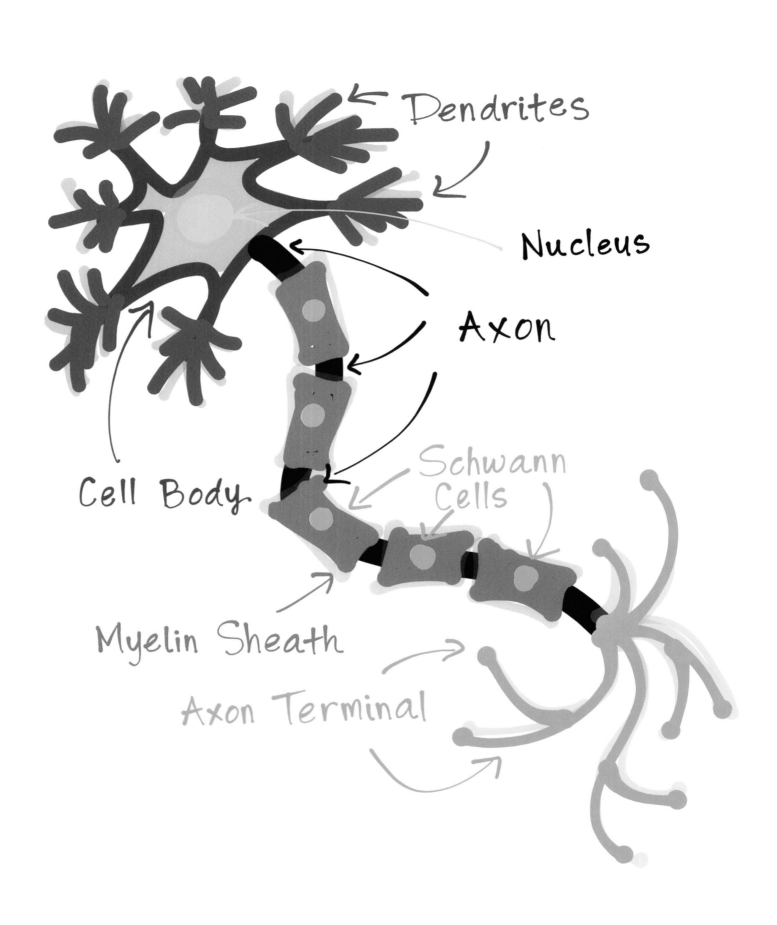

Dendrites

Nucleus

Axon

Cell Body

Schwann Cells

Myelin Sheath

Axon Terminal

Axon Terminal

The axon terminal receives the message from the axon.

The axon terminal sends the message to the next neuron's dendrites.

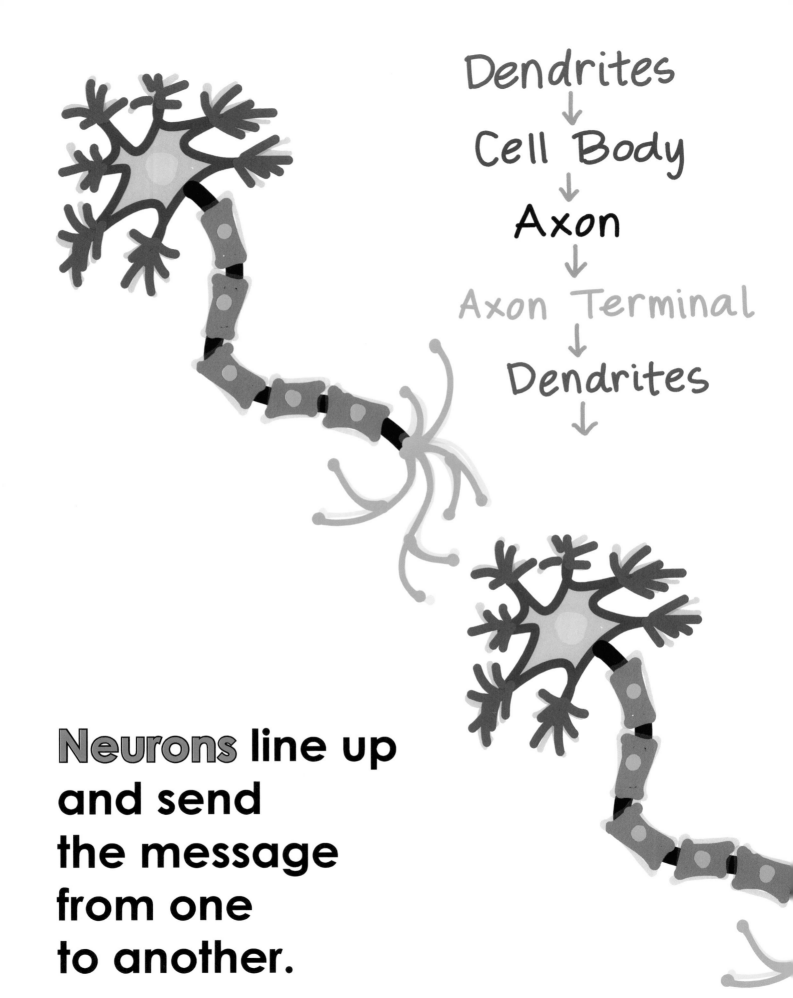

Dendrites
↓
Cell Body
↓
Axon
↓
Axon Terminal
↓
Dendrites
↓

Neurons line up and send the message from one to another.

This is how a message can get from your hand or foot all the way to your brain.

"Remember this" REVIEW

Frontal=**?**
Motor=**?**
Somatosensory=**?**
Parietal=**?**
Occipital=**?**
Cerebellum=**?**
Brain Stem=**?**
Temporal=**?**

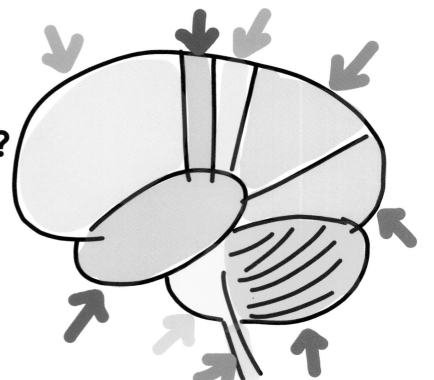

Neurons are the connections between your **brain** and your body.

Messages are sent from the **dendrites** to the **cell body**, then down the **axon** to the **axon terminal**, and to the next **neuron**.

The spinal cord is made of bundles of neurons.

It is the pathway for messages between the brain and body.

The spinal cord is protected by bumpy bones called vertebrae.

You are a Neurology expert!

yay science!

The next book coming in the Super Smart Science Series is:
ASTRONOMY: THE SOLAR SYSTEM

Do you have a science topic suggestion for the Super Smart Science Series**?**
Let us know at www.Facebook.com/SuperSmartScienceSeries.

Submit photos of your family reading our series **to be featured on our website!**

Free activities and worksheets **available online at www.SuperSmartScienceSeries.com.**